teri | THE ENERGY AND RESOURCES INSTITUTE
Creating Innovative Solutions for a Sustainable Future

Government of the Netherlands

Reclaiming the Blue Earth

Neha
•
Pragya Gaur

Foreword

Imagine if the world's water quantity could be poured into a one-litre bottle—only half a teaspoon could be used as drinking water. The rest is too brackish, polluted or difficult to reach.

We must treat our water system as precious and at the same time protect citizens from the threat of water. In the Netherlands, we live every day with water as most of our country is far below sea level. Over 1000 years, this has been a struggle and a fight against water.

Nowadays, we have developed a water system in which we can drink water from the tap and developed sophisticated dikes and barriers to protect us against the sea and rivers.

The solutions are actually very simple. It is just a matter of good governance between cities, regions and states, awareness amongst citizens, and finally a bit of technology. However, in many countries this is very complex for social, economic, and cultural reasons.

In the 1970s, the water quality of rivers, lakes and canals in the Netherlands was extremely bad. By making agreements with other countries and cooperation with all relevant parties, we now benefit enormously with clean water in terms of economy, culture, and recreation. Water gave us a lot of wealth and joy.

Still, we cannot take that for granted. New challenges come up like droughts, floods, sea-level rise, and water pollution by chemicals. These challenges do not only count for the Netherlands. They count for the whole planet.

It is extremely important to connect people with all water-related issues, whether it is on a better understanding of the water system or cleaning the river banks together with the community.

Young people can be the real water ambassadors in this regard and I am proud that the Dutch government in collaboration with TERI could launch the "LIVE with Water" education programme. This fits in a long history of collaboration in the field of water between India and the Netherlands.

This book is one of the many steps towards a healthy water system, which will finally create lots of prosperity and joy for the people of India.

Bart Jeroen Bierens

Senior Water Advisor at Netherlands Enterprise Agency

Message

Water is extremely important in our daily lives. Not only we need water to survive, but we also need water for washing and sanitation, for cooking, growing food, raising animals or for producing different goods, to name a few. However, using water without proper treatment afterwards leads to water loss and water pollution. And this has significant negative impact on our lives.

Water pollution used to be a big problem in the Netherlands and other European countries in the 1960s and 1970s particularly. It manifested itself from time to time in the form of massive fish mortality, bad odours and very polluted lake beds, riverbeds and sea beds. The government tackled these problems by imposing far-reaching limitations on discharges by companies and improving the treatment of (waste) water and by collaborating on water management with the neighbouring countries. As a result of these efforts, and due to the development of phosphate-free detergents, the water quality improved very quickly.

However, more can be done, not only in Europe, but also worldwide. With the growing demand for clean and safe water, it is important to raise and spread awareness about wastewater-related issues, and the way we can address them. India and the Netherlands have a long-standing relationship in the bilateral water cooperation, addressing not only the problem of wastewater, but also water conservation, improving water quality and flood management. The Netherlands and India are engaged in a broad range of water-related projects embracing the Sustainable Development Goals. This provides a solid foundation for bolstering and accelerating cooperation.

I am happy that India and the Netherlands joined their forces in the 'LIVE with Water' educational project. The thought of initiating the project goes back to the LOTUS-HR project (The Local Treatment of Urban Sewage Streams for Healthy Reuse) initiated by The Government of Netherlands and TERI at Sun Dial Park to treat 10,000 litres per day of sewage water from the Barapullah drain, Delhi.

The Indo-Dutch 'LIVE with Water' project identified and addressed the water curriculum gap, onboarded several schools and organized different sessions, workshops and activities for both the students and teachers, with an aim to raise awareness of the importance of water use and reuse, and to give them concrete tools for spreading the knowledge.

The book 'Reclaiming the Blue Planet' that you are reading is a very informative and useful product of the project. It aims to teach us how to change the way we treat water. Read it yourself and pass it to your family and friends to raise the awareness and share the knowledge. And do value water. Because it is the most important element of our lives.

Marten van den Berg

Ambassador to The Netherlands in New Delhi

Message

Our planet is mostly covered with water. It has bodies that comprise a significant portion of water. Humans' demand for sustainable supply of water is undeniable. However, humans have not always been proactive in responsibly using resources that are finite, depleting, and irremediable. While the idea of water reclamation is not particularly new, changing consumption patterns, global climate change, increased population, and land use have further strained water resources and supply challenges. A problem that has varied manifestations and is progressive requires a nuanced approach.

This book is part of a project titled 'LIVE with Water', which is an Indo-Dutch water programme. The Energy and Resource Institute (TERI) in collaboration with the Norwegian Government commenced this programme to generate awareness among students of grades 6–8 on treated wastewater use. This initiative, notionally, enables the acknowledgement of an individual agency. Engaging the youth becomes imperative in such aspects, as it is their outlook, perspective, and attitude that can sustain any and every effort made in the direction of a sustainable and aware future.

This document is a curated convergence of a curriculum assessment, consultation with stakeholders, content development, expert evaluation, and effectual procurement of materials. The book is purposive of encapsulating all suggested deliverables and addressing the identified gaps while effectively fostering critical thinking and problem-solving abilities.

Universal access to safe and affordable drinking water for all, a goal that the SDG's are attempting to achieve by 2030, is propped on the conscience and cognizance of the citizens who are school children today but will be the youth of tomorrow. Water management is a challenging subject due to competing sectoral demands. This has both pros and cons. Since a large number of agents are involved, water-related problems can be managed collectively. However, it may be a shortcoming, because it is difficult to collaborate, build a shared vision, and form a consensus on the goals and allocations.

This book is the result of cumulative experiences and efforts. The experiences provide an extensive and integrated perspective of the technical, conceptual, and organizational aspects for sustainably implementing water-reuse practices across generations. It is a document that reimagines the constructs of 'education', with respect to sustainable and aware actions. It strives to impart knowledge, enable contextual comprehension, and promote enduring practices and attitudes, leading to a more secure and sustainable future.

This book is a step in the direction of an effortlessly conscientious future, a life and living with water. A step that we hope you take along with us.

Vibha Dhawan

Director-General, TERI

Preface

If there is magic on this planet, it is contained in water. Without water, there will be no hues of green, red, brown, and yellow around us as there will be no life. Water is an essential resource but it is finite, though renewable. With the growing population, there has been a never-ending demand for water. It is estimated that more than 3 billion people worldwide do not have access to drinking water. By 2025, approximately 1.8 billion people will be living in regions or countries facing a situation of absolute water scarcity. Occurrence of water scarcity could be either lack of quality of water or its quantity. Worldwide, agriculture is the largest consumer of water, followed by the industrial sector, and then the household. In developing countries, the share used by agriculture and households is higher than that in developed countries. Emerging trends indicate that in the near future, the main constraint to agricultural production may be the non-availability of water, not land. More than five million people, most of them children, die every year from illnesses caused by drinking poor quality water. With competing sectoral demands of water, its management has been a complex issue—this aspect has its own strengths and weaknesses. Strength because water-related problems can be solved on a collective basis as a multitude of stakeholders are involved. A weakness because it is indeed difficult to work collectively, develop a common vision, and agree on priorities and allocations. One of the United Nations Sustainable Development Goals (SDG: 6, Clean Water and Sanitation) by 2030 is to achieve universal and equitable access to safe and affordable drinking water for all.

In the light of the mentioned facts, it becomes essential to take action now for our future generations before it is too late to address the issue. 'Reclaiming the Blue Earth' aims to generate awareness and develop consciousness about the use, reuse, and reclamation of water. The book also makes the readers realize their own strength and role; change their thoughts, perspectives, and attitude towards the use and misuse of water. Contents are developed to sensitize people towards the use and reclamation of wastewater through critical thinking and problem-solving. The text has been supplemented with simple solutions for effectively dealing with potential problems related to water at homes and in community, at large. Let us take a step forward to save water for our generations to come.

Acknowledgements

To ensure a habitable future—one that is no longer a choice, but an acknowledged need—is our responsibility. Collective effort and personal responsibility is a prerequisite to all forms of sustained impact and effective action. Globally, there is an increase in united efforts and action towards a liveable earth and sustainable future. The realization of collaborative efforts culminating into an aware sustained way of life is thriving, with potential that can be channelized to improve our future. *Reclaiming the Blue Planet* book is a cumulative aggregate of varied expertise, emerging from effective collaboration and contributions, purposed to contextualize the efforts, notions and perceptions, requisite for us to "LIVE with Water".

The content of *Reclaiming the Blue Planet* is a harbinger of hope and resilience as it very rightly targets the future generation by apprising and sensitizing them to the prevailing water situation and facilitating their participation to conserve and re-use wastewater. The book is designed and developed with appropriate need-based inputs in consultation with experts with five distinctive processes- curriculum review, consultative meetings, development of content, technical review, and design/editing. Each of these methods and practices have been diligently carried out to ascertain the development of content that is universal in its access and impact. The novelty of this book is in its reach, sufficing as an educational manual for students and reference material for teachers, as well as a library book for an intrigued reader. It is now, more than ever before, that we should reach out for a book, such as *Reclaiming the Blue Planet*.

It is imperative to convey appreciation and acknowledgment to the numerous people who have participated and contributed to the manifestation of this content.

Our special thanks to Mr Bart Jeroen Bierens and Dr Fernanda van der Velde for the stimulating suggestions and encouragement to make this book possible.

For the process of curriculum review and identification of present gaps, the engagement of Ms Zoya Khan and Ms Muskaan Mascharak is valued. Additionally, Ms Saltanat M Kazi and Ms Monmi Barua are acknowledged for their efforts in carrying out consultative sessions with teachers from diverse fields and various school boards. A special gratitude to eminent educators—Ms Poonam Virmani, Ms Sudha Rani, and Ms Marina Sanyal—for being a support pillar and contribute during consultative sessions. This process proved to be particularly helpful in effective identification of deliverables. After the review and consultation, the process of developing the content followed, for which Ms Samhitha D notably contributed and Ms Saltanat M Kazi ably seconded and aided the process of content development.

Subsequently, it is imperative for any content to undergo a rigorous technical review for an ensured check on quality. Therefore, we express utmost appreciation and gratitude to Prof. Arun Kansal to have lent his expertise and validated the content for *Reclaiming the Blue Planet*.

Finally, we thank the publishing team of TERI comprising the editorial and design team. Mr Santosh Gautam and Mr Vijay Nipane presented the content effectively using interactive design elements, lively illustrations, graphics, and visuals.

The culmination of all our efforts is only realized with a chance to be received well by our readers.

Hoping for students with awareness, proactive educators, and responsible readers, to form a society that thinks for the future in the present.

Contents

Chapter 1

Water in Our World

Do you ever wonder why scientists are busy in looking for water on other planets? Because it is a unique natural resource amongst all sources available on earth. Life is not possible without water, and all living things need water in some form to survive. People can go weeks without food but can live only a few days without water. About 75% of the earth's surface is covered with water, but how much can we use?

Out of all the water available on the earth, 97% is saline and is in oceans, 3% of water is freshwater available in rivers, streams, and glaciers. Let us have a closer look at the distribution of water on the earth.

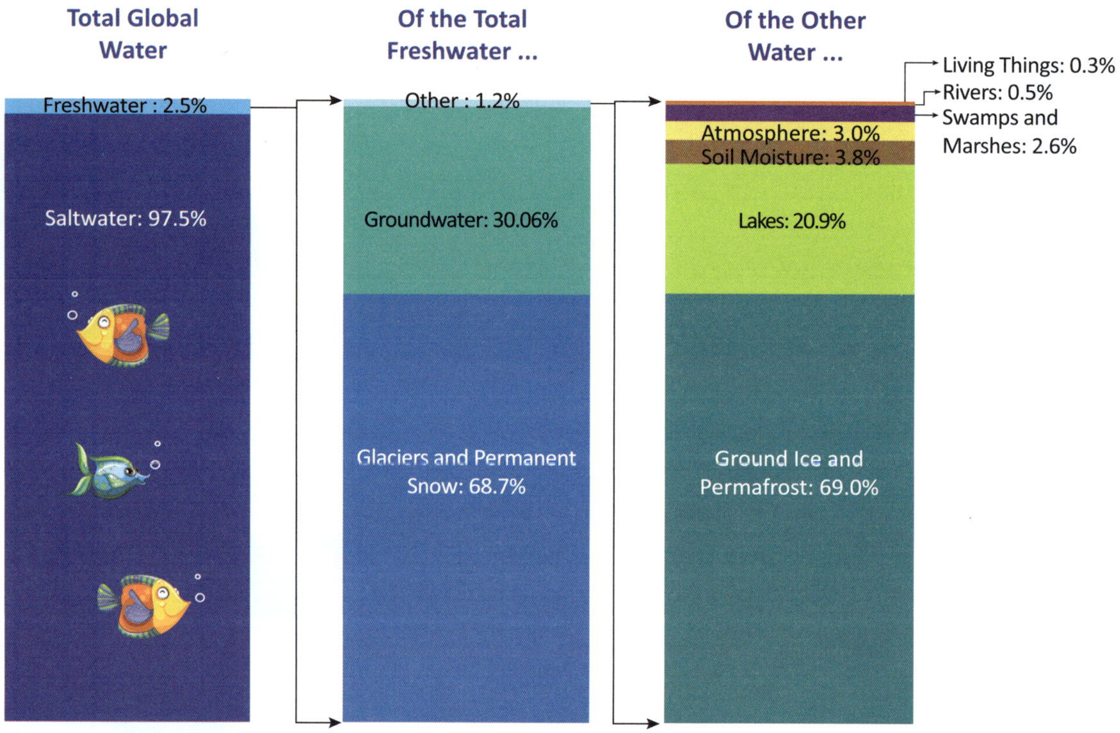

Activity

Interpret the graphic organizer (given above) and answer the following questions:

1. Where is most of the earth's water present?
2. Why can't we use salt water?
3. How much water is available for human use?
4. Do you think available water is equally distributed around the Earth?

The Recycling of Water in Nature: The Water Cycle

Water never sits still. Water supply constantly moves from one place to another and from one form to another through the water cycle. The water cycle, also called the hydrological cycle, explains how water moves from the land into the atmosphere and then back to the ground and oceans. It consists of four stages: (i) evaporation, (ii) condensation, (iii) precipitation, and (iv) collection. This system is a continuous exchange of moisture between the oceans, the atmosphere, and the land. The process derives its energy from the sun, heating all the earth's water and making it evaporate.

DID YOU KNOW?

The amount of water in the atmosphere at any moment in time is only 12,900 cubic kilometres a minute fraction of earth's total water supply: if it were to completely rain out, atmospheric moisture would cover the earth's surface to a depth of only 2.5 centimetres. However, far more water in fact, some 4,95,000 cubic kilometres of it, is cycled through the atmosphere every year. It is as if the entire amount of water in the air was removed and replenished nearly 40 times a year.

Activity Write the processes that drive the water cycle on the earth in the given graphic.

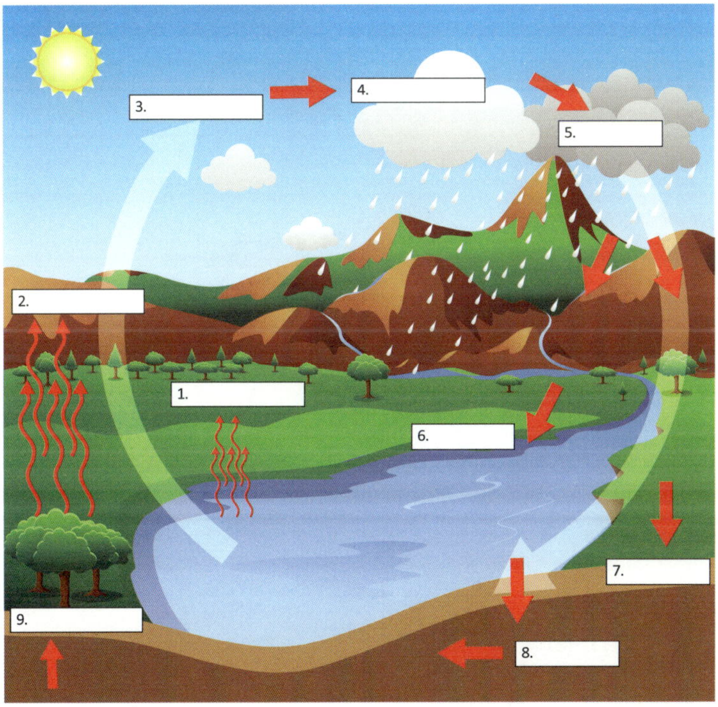

Answer key

1. Evaporation, 2. Transpiration, 3. Condensation, 4. Transport, 5. Precipitation, 6. Runoff, 7. Infiltration, 8. Groundwater, 9. Plant uptake.

The Water Cycle and Climate Change

The water cycle is vital to life on earth and weather and climate across the planet. However, even slightest changes can alter its cycle. Owing to global warming and climate change, the temperatures rise and the water cycle speeds up, weather becomes all the more unpredictable, and the conditions more and more extreme.

The water evaporates from the land and sea, which eventually returns to earth as rain and snow. Climate change increases the rate of evaporation as air temperatures increase. One might say that the increased evaporation can be balanced by higher precipitation or rainfall. However, it is not caused because of wind currents and weather patterns on earth, water that evaporates rarely falls at the same place. It collects, forms clouds, and is blown across plains or mountains before it falls again which causes some areas to experience stronger storms, others experience more dry air and even drought. Hence, the slight disturbance can lead to more intense changes in the pattern of rainfall and precipitation. One expected effect of climate change will be an increase in precipitation intensity: a larger proportion of rain will fall in a shorter amount of time than it has in the past. This will lead to rainstorms, causing major problems such as extreme flooding in coastal communities around the world.

Impact of Climate Change on Weather

The risk of the occurrence of droughts and floods is increased due to changes in water cycle

- Due to higher temperatures, there is increased evaporation from land and oceans into the atmosphere.
- Warmer air holds more water vapour, which can lead to intense rainstorms.
- There is increased risk of flooding due to intense rainstorms. Most of the water runs off into rivers and streams, doing little to dampen soil.
- Combined with increased temperatures, increases the likelihood of droughts.

Activity

The map here represents the predicted climate model of precipitation change by end of the 21st century. dark green represents areas where climate models predict an increase in intensity while purple represents a predicted decrease. (Map adapted from the IPCC Fourth Assessment Report.)

What changes in rainfall patterns in India can you predict looking at the map?

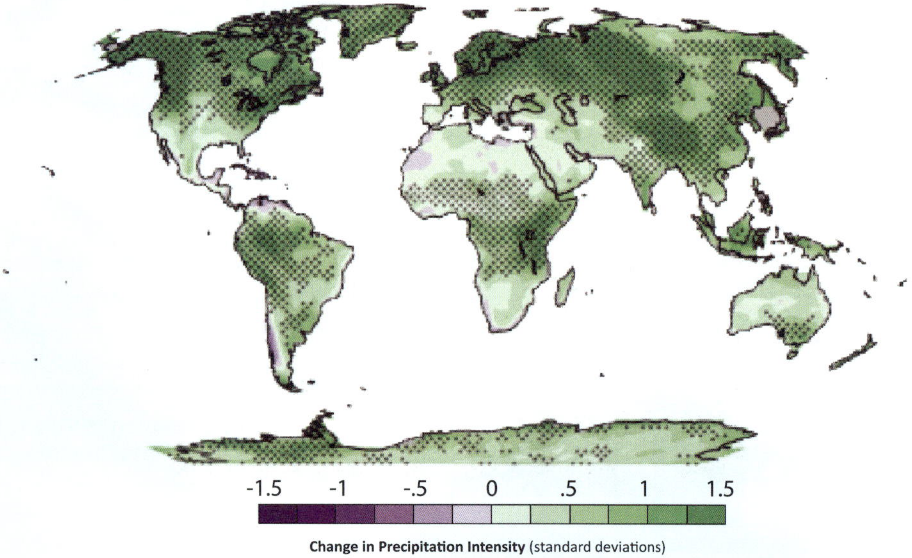

Change in rainfall pattern

TASK

Now that we know about the water cycle and predicted changes in the water cycle by end of the century. Make a cartoon strip depicting the journey of water. Think about the hurdles it will face in today's scenario, take into account air pollution, water pollution, global warming, and climate change. What might a drop of water feel after facing these challenges on its journey?

Post your entry on Facebook page. Use QR code as on page no: Insert page no 44

The Geographical Distribution of Freshwater Around the World

Even though the water cycle keeps the water in motion over the entire earth, the geographical distribution of freshwater is highly uneven. The available water is not evenly distributed around the planet, and neither are the people.

World's 70% freshwater, as glacial ice, is trapped, in Greenland and Antarctica. Most of these resources are located far from human habitation and are not readily accessible for human use. Groundwater is the most abundant and readily available source of freshwater, followed by lakes, reservoirs, rivers, and wetlands. Analysis indicates that groundwater represents over 90% of the world's readily available freshwater resource. About 1.5 billion people depend on groundwater for their drinking water supply. A region's water population balance largely is determined by the difference between its proportion of the world's available water and its proportion of its population. Let us examine the freshwater distribution, through a case study, across continents and the distribution of people to understand how much water is available for use in a particular region.

Case study

The infographics below represent the indicative distribution of fresh water in the seven continents, and populations for 2021. Study the data and answer the questions that follow.

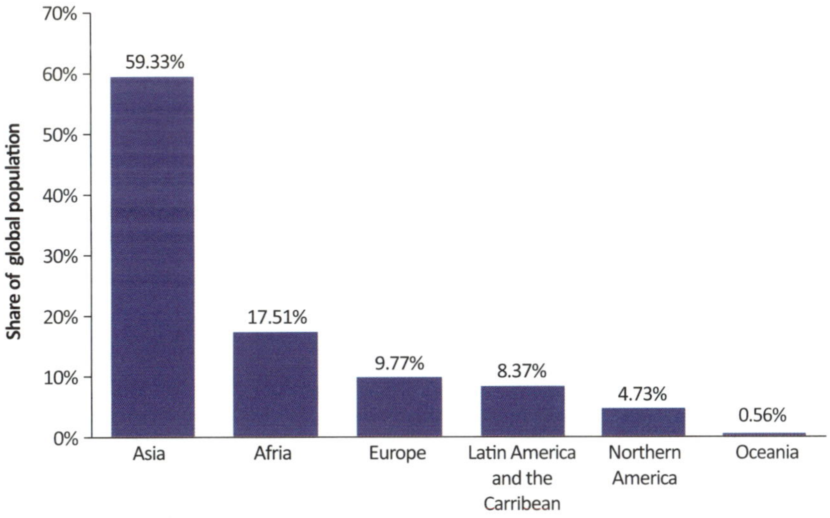

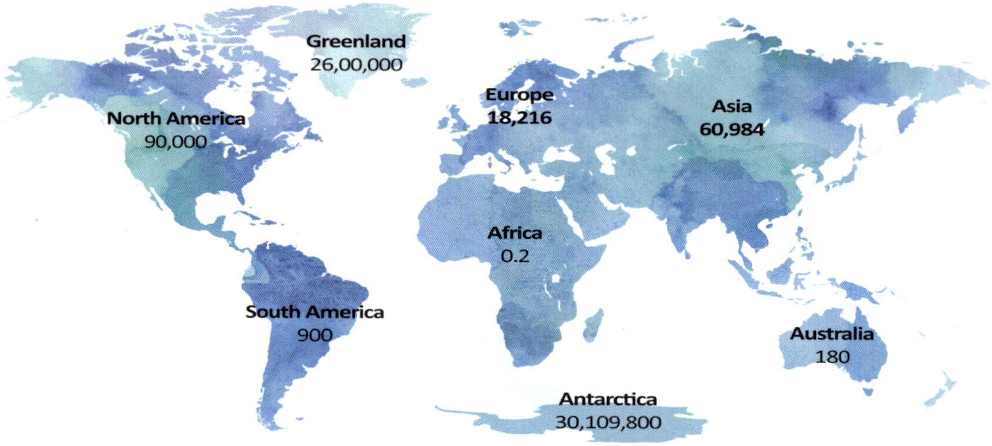

	Glaciers and permanent ice caps (km^3)	Wetlands, large lakes, reservoirs and rivers (km^3)	Groundwater (km^3)
North America	90,000	27,003	4,300,000
Greenland	26,00,000	-	-
South America	900	?	3,000,000
Europe	18,216	2,529	1,600,000
Africa	0.2	31,776	5,500,000
Asia	60,984	30,622	7,800,000
Australia	180	221	1,200,000
Antarctica	30,109,800	-	-

On the basis of information contained in the graph and data given in table answer the following questions.

1. Given the population of the Asian continent and the amount of water present in lakes and reservoirs, what can you predict about the water availability in the region by the year 2030?
2. In Asia, most freshwater is spent on agriculture to feed the massive population. Considering the limited freshwater availability, how do you think the Asian continent will provide its people with the growing demand for food and water?
3. Given the current scenario where water is getting scarce, give two suggestions on how can a continent or a country conserve its water resources.

The Alarming Numbers

According to *The United Nations World Water Development Report 2018*, the world population is expected to increase from 7.7 billion in 2017 to between 9.4 and 10.2 billion by 2050, and the global gross domestic product is expected to increase by 2.5 times in the same period. With the rapid population and economic growth, the report estimated that the global water demand will increase from about 4,600,000 mcm (million cubic metre) per year in 2017 by 20% - 30% to between 5,500,000 to 6,000,000 mcm per year by 2050.

How Is Water Used Around the World?

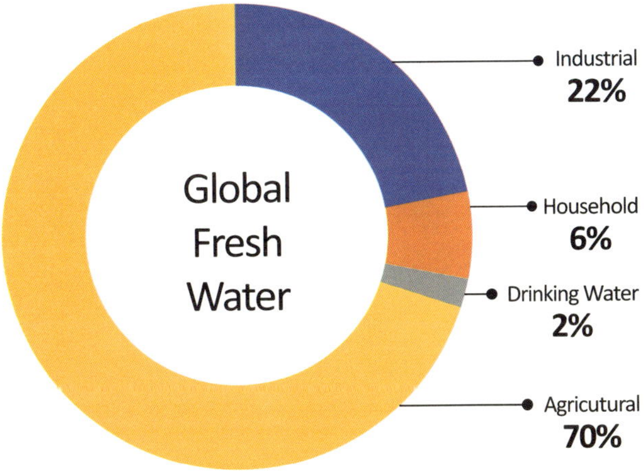

The amount of water used on a global scale has increased rapidly in recent years. This increased demand is due to population growth and greater per capita consumption of water. The rate of increase in industrialized countries is the lowest; most of the increase is in the developing world, with a much lower per capita water use: 70% of global water use is for agriculture, 22% for industry, and 8% for domestic purposes.

This division has a considerable regional variation, with agriculture being the more water-demanding and using 85% of water in some regions such as Asia and Africa while industries from Europe and the USA also use a more significant proportion of water.

Considering the given statistics and the dwindling supply of water due to factors triggered by climate change such as fluctuations in precipitation, reducing snowpacks, drying out surface resources, agricultural regions around the globe have been subject to extensive and increasing water constraints. These and other extreme weather events, such as floods or tropical storms, are also expected to be more frequent in the coming years. Major droughts in Chile and the USA have affected agricultural production while diminishing surface and groundwater reserves. An OECD (Organisation for Economic Cooperation and Development) assessment of future water risk hotspots projects that Northeast China, Northwest India, and the Southwest United States will be among the most severely affected regions, with domestic and global consequences. So, what can policymakers and farmers do to minimize water use? The governments should act at the farm, watershed, and national levels to:

- Strengthen and enforce existing water regulations.
- Create incentives for farmers to improve their water use and better manage the use of polluting agricultural inputs.
- Remove policies that support excessive use of water and polluting activities.

The farmers can use various techniques for sustainable water supply in agriculture that include organic farming practices, which limit substances that would contaminate water, efficient water delivery, micro-irrigation systems, mobile technology, precision farming, adaptive water-lifting technologies, zero tillage, rainwater harvesting, runoff farming, and drip irrigation.

TASK

Can farmers reuse or reclaim wastewater for irrigation? Find out.

If we see the current distribution of water resources, on a global scale, water is not scarce, but if we observe it locally on a continental or national scale water is often scarce, and with increasing demand is likely to be more so in the future due to rise in global population and increased demands of agriculture and industries.

Case Study

The infographic depicts the water usage trends across the world approximately five years back. Study the given infographic and answer the questions given below.

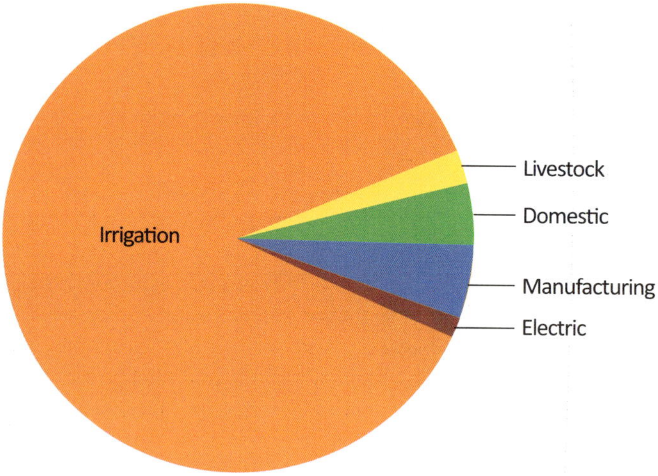

Pattern of global consumption of water

a. What does the global consumption pattern of water indicate?

b. Given that the world population is growing daily, how would water usage change by the next five years?

c. What changes you might suggest to policymakers and people working in the agriculture sector to minimize water use? Suggest two technologies that farmers can use to reduce water usage without compromising the yield of the crops.

Water Availability in India

India is a populous nation with 18% of the world's population. It has an extensive river system with 11 big rivers and their tributaries. However, it is only 4% of the water resources. These water resources are facing severe threats due to various natural and anthropogenic activities. Now, a significant share of the water goes into agriculture to support the needs of the growing population. With the steady increase in population, there is tremendous stress on water too, which is incessantly increasing.

Case Study

The graph below compares per capita or person water availability over 50 years.

Think about the trend projected in the graph and share the factors contributing to water scarcity on the project's Facebook page. Use QR code as on page no: 44

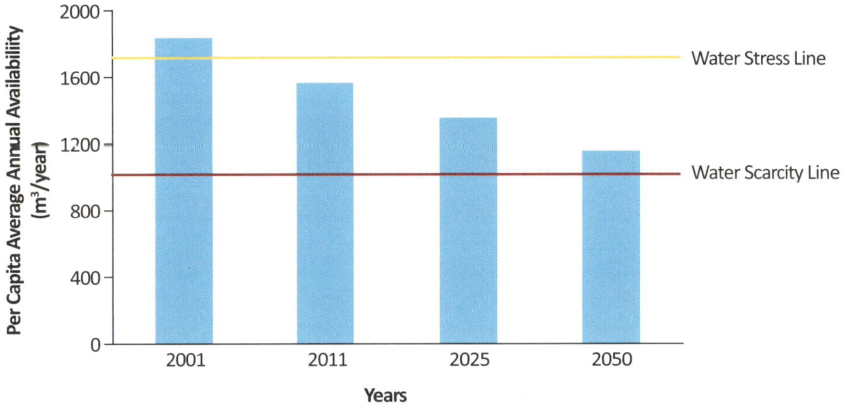

Source: https://nwa.mah.nic.in/sdmc/availability/sufficiency.htm

Reflect

Observe and analyse the pictures given below. Do you think we are using our water resources wisely?

Given that water is a precious resource, the images above are evidences of our carelessness and suggest that we do not use it wisely. What actions will you take, or what will you tell each and every person responsible for these scenarios?

Let us investigate further the need and usage of water in our day-to-day life.

Chapter 2

Water in Our Life

Water has always influenced where human development occurs. Humans need water for so many purposes that it affects how an area develops tangibly. Ideally, people would move into areas with adequate water supplies. Most ancient civilizations were established near the rivers. Humans use water six times today as they did 100 years ago. People living in developed countries use a far more significant proportion of the world's water than people in less developed or developing countries. How do people use all this water? What are the various sources of water? Are they natural or man-made? Explain that people use water in many different ways.

Identify the names of the nearest water resources available around their homes/schools with the help of India's topographic map, categorize them as natural and human made resources.

Natural Water sources	Human made storages
Rivers	Dams
Lakes	Swimming pools
Ponds, Springs	Reservoirs

Activity

We use water for drinking, cooking, washing, cleaning, gardening, etc. Given below are few activities that need water. Predict the amount of water that we generally consume or use in these activities. Once done, compare and see if your prediction was correct or closer to the average amount mentioned in the answer key.

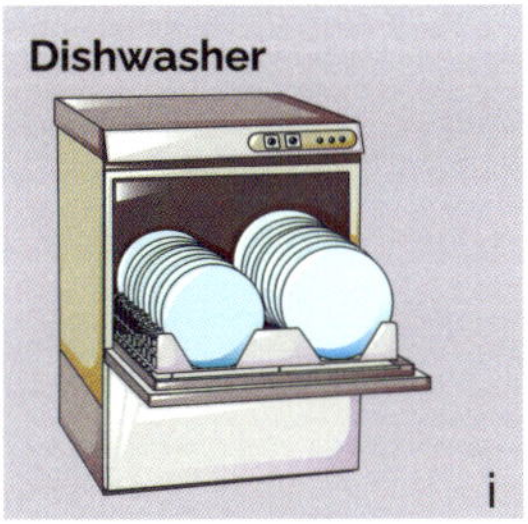

Answer key

a. 0.01–1 L; b. 1–2 L; c. 1–5 L; d. 1–3 L; e. 5–10 L; f. 50–150 L; g. 1–40 L; h. 30–100 L; i. 5–200 L; j. 5–200 L; k. 1–17 L

Activity

Take a note of your household behaviour over one week. Take a log of water used and wasted while doing the chores. Use this table format to help you keep a diary of how much water you use at home.

Activity	Flushing Toilet	Washing to maintain hygiene	Drinking water/hot and cold drinks	Washing clothes using machine	Hand washing dishes	Dishwasher	Shower	Bath	Brushing teeth	Washing Car
Amount used	6 litres per flush	6 litres	0.5 litre per drink	50 litres per wash	12 litres	14 litres	150 litres per power shower 40 litres for normal	80 litres	6 litres	35 litres per bucket
Monday										
Tuesday										
Wednesday										
Thursday										
Friday										
Saturday										
Sunday										

In reference to the data collected answer the following questions.

a. How much water do you use every day for bathing?

b. Do you take a power shower or use a bucket?

c. Do you keep the tap running while brushing teeth? How much do you think is waste?

d. Are you using water wisely enough?

e. How much water do we need per day for living comfortably? Do we use much more water than needed? If so, why?

Activity

Due to water shortage and scarcity, a city has decided to limit water for each individual. Each person is allowed to spend only 20 litres of water. How will you spend your 20 litres of water a day? Use the following diamond nine priority matrix template to list how you will use water. List the high priority activities in the top diamonds and least important at the bottom. Write other activities of medium priority in middle-level boxes as indicated on the template.

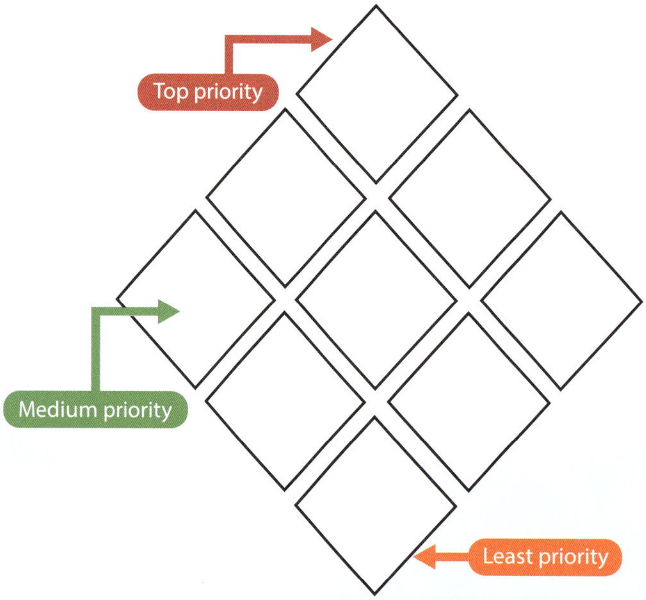

Reflect

Are we using too much water and wasting it too? Can you reuse some of this water?

Activity: Stop the Drip Drop!

Materials required: Stop-watch, measuring container, notebook and pencil

1. Choose a tap that is frequently used at home or school. Before starting the activity, predict how much water would be collected in the container for a specified time and ask to note this in the notebook. Now, let the tap drip where the water falls in the form of droplets. Your task is to keep the measuring container and time how long it took to fill up.
2. For another variation, increase the water flow from the taps where a thin thread of water is falling instead of water droplets. Again, time the collection of this water in the water container.
3. Now, calculate the amount of water lost due to leakages (both options) for 24 hours and then for a month.

In the case of Delhi, it is estimated that of the total water supplied,

40% water

is lost due to water distribution leakages such as poor piped network, breakdown /damage of pipelines.

Record and Reflect

1. How close was your initial estimation to the findings?
2. Discuss how this water can be reused.
3. What will be your response if you see dripping water taps?

Post your entry on Facebook page. Use QR code as on page no: 44

Chapter 3

Water Footprints: Direct and Virtual

When we think of our water use, we only consider the water used in drinking, washing, cooking, bathing, and other routine household chores. If someone tells you that we eat, wear and carry water with us and have water in things present all around us. Will you agree? We might not; however, the truth is that the food we eat, clothes we wear, and the leather or plastic bag we carry all need water to be manufactured and transported to us.

This indirect water is called virtual water. For example, a can of cola contains 0.35 litres of water, yet it requires an average of 200 litres to grow and process the sugar contained in the can. Similarly, it takes 2,900 litres to 'grow' a cotton shirt and 8,000 litres to produce a pair of leather shoes, i.e., the amount of water required to grow feed, support a cow, and process its skin into leather. For example, if you are washing your vehicle, you might be using direct water whereas, when you are feeding your hen, that feed also requires a lot of water to produce, and the hen provides eggs or even become food that people eat. So, when we eat chickens or eggs, we consume water virtually (i.e., indirectly).

Water Footprint

We can estimate our total water footprint if we total up all of the virtual water embedded in everything we use and eat.

Therefore, the total water demand of a country or an individual is not only the amount of water withdrawn from the water bodies. It also includes the global water sources that are required to produce imported products. The water footprint consists of two parts: domestic water resources and water use outside the country's borders. Thus, the water footprint includes the water being withdrawn from the surface and groundwater and soil water use (in agricultural production).

A water footprint is the total virtual water content of products consumed by an individual, business, town, city, or country.

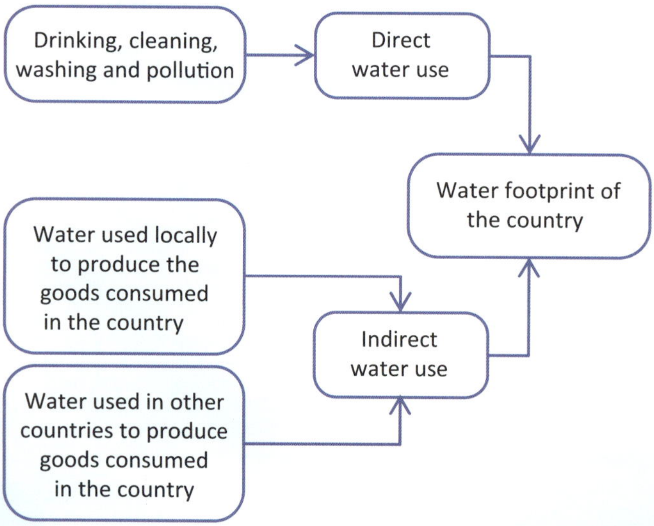

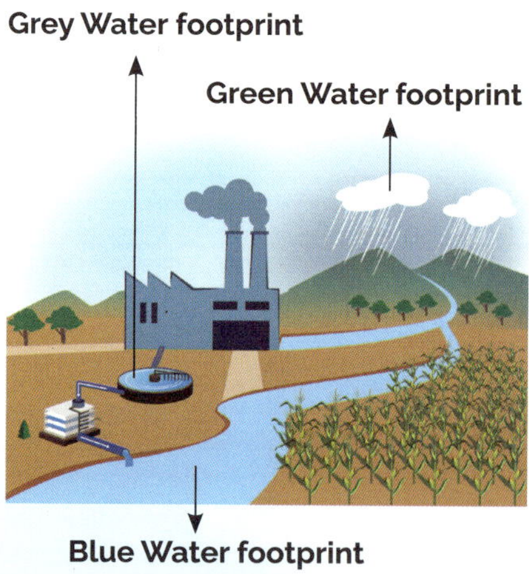

Water footprints of some common products we use are as follows:

Smartphone

910 L

1 A4 sheet of paper

10 L

Polyester jacket

18,000 L

Cotton t-shirt

100 L

Leather bag

4200 L

A pair of jeans

800 L

Leather shoes

5000 L

1 apple

125 L

1 orange

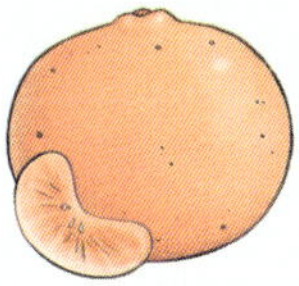

80 L

1 tomato

50 L

1 slice of bread

160 L

100 g cheese

318 L

1 beef hamburger

2400 L

1 pizza margherita

1260 L

1 glass of milk

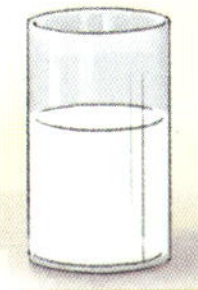

255 L

1 egg

196 L

1 cup coffee

132 L

Our food and clothing cannot be made without a great deal of water. Freshwater ecosystems support agriculture to produce cotton and biofuel; regulate water flows; purify wastewater and detoxify wastes; control climate; protect from storms; reduce erosion; and offer cultural benefits, including significant aesthetic, educational, and spiritual benefits. We are withdrawing freshwater in quantities and at rates greater than nature's ability to 'renew'. One such example includes the Aral Sea, which has come to epitomize the devastating economic and ecological effects of water mismanagement more than any other water body in the world.

Activity

Guess the Water Footprint

Let us analyse our water footprint by predicting some common foods that we consume each day. Let us estimate the water footprint of the items given in the following table and compare our estimate with the actual numbers.

1 — **100 g chocolate**

2 — **100 g bread**

3 — **100 g pasta (uncooked)**

4 — **100 g chicken (cooked)**

5 — **250 mL tea (without milk and sugar)**

6 — **125 mL coffee (without milk and sugar)**

Answer key
1. 1700 L, 2. 130 L, 3. 141 L, 4. 433 L, 5. 123 L, 6. 306 L

Based on your findings about water footprint, think and answer the following questions.

1. How could each of us save water by making small changes to the items we consume?
2. Which water footprint values did you find most surprising? Did any shock you?
3. What types of food do you think typically have high water footprints and why?
4. What factors do you think might contribute to the water footprint of a food?
5. Why might the water footprints of similar items (e.g., coffee and tea) be very different?
6. If you were required to reduce your water footprint, what effect would this have on you and your family?
7. One-third of the world's food production ends up as waste. How do you think we should reduce our food waste to decrease the amount of water wasted?

Activity

Lunch Box Water Footprints

What did you have for lunch today? A subway sandwich or stir-fry rice with vegetables or chicken dumplings. Every day we relish various dishes, try different cuisines, and prepare exotic meals. Have you ever thought how much water these use to get cooked, to grow, and to reach us using means of transport such as airplanes, trains, trucks or vans? What might be the water footprint of your lunch meal? Give it a number.

Now, calculate the water footprint for your lunch box. Pick three of your favorite lunch meals. Research the water footprint values of food items and calculate their total water footprint. Post your entry on Facebook page. Use QR code as on page no: 44

a. Which lunch items have the highest water footprint, and which has the least?

b. Which item from your lunch box will be a sustainable option over the years, considering its water footprint?

c. How might you replace your lunch box with better options?

d. What other ways can you think of to reduce your food water footprint? (For example, change from coffee to tea and from beef to chicken.)

TASK

Design a lunch meal of your choice with a footprint of not more than 300 litres.

Reflect

Write a paragraph describing your assessment of the value of water footprint analysis. What have you learned from it? How will you use this information to reduce your water footprint?

Post your entry on Facebook page. Use QR code as on page no: 44

Water Footprint and Sustainability

We tend to think first about reducing our direct water consumption. Still, virtual water is the most considerable portion of a person's water use because globally water for food is the most significant part of the water most people use.

Some water advocates encourage people to eat less meat or switch to a vegetarian diet. Looking at the values from people eating a meat-based diet and those eating a vegetarian, vegan, or plant-based diet without much meat, what can you conclude about water usage and dietary connections?

Stop and Think

How water conservation efforts often focus on more sustainable production or finding ways to decrease the amount of water we use to produce goods and services?

Why might it also be necessary to consider more sustainable consumption or the number and type of products we consume and where and how they're produced?

Chapter 4

How Blue Is the Blue Planet: Water Scarcity?

According to the Global Risks Report 2017, 4 billion people currently experience water shortages for at least one month of the year. The World Economic Forum places water crises third in its list of top risks as about 36 countries currently face incredibly high levels of water stress. A water crisis occurs when the available potable, unpolluted water within a region is less than that region's demand. This can be the result of two mechanisms.

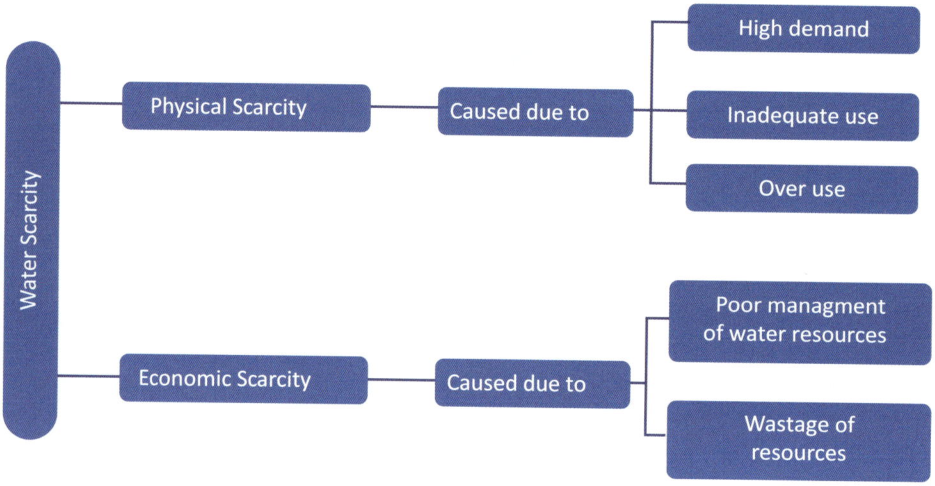

The most commonly used method of defining water scarcity is the 'water stress index', which measures the amount of renewable freshwater available for each person each year, considering the total water resources available and the population each year.

According to this index, if the total renewable water available in a country is below 1700 m^3 per person per year, that country is said to be experiencing water stress (below 1000 m^3 – is experiencing scarcity; below 500 m^3 – absolute water scarcity).

Case Study

To access the world water interactive map, use the link https://worldwater.io. Analyse the map and answer the following questions.

a. Which is the driest month in India?

b. Compare the water stress, water scarcity, and absolute scarcity numbers of India, China, and Afghanistan. What do the numbers suggest?

c. After analysing the data of various countries on water scarcity, comment on the world's water crisis situation.

Case Study

Analyse the map given below and answer the questions that follow.

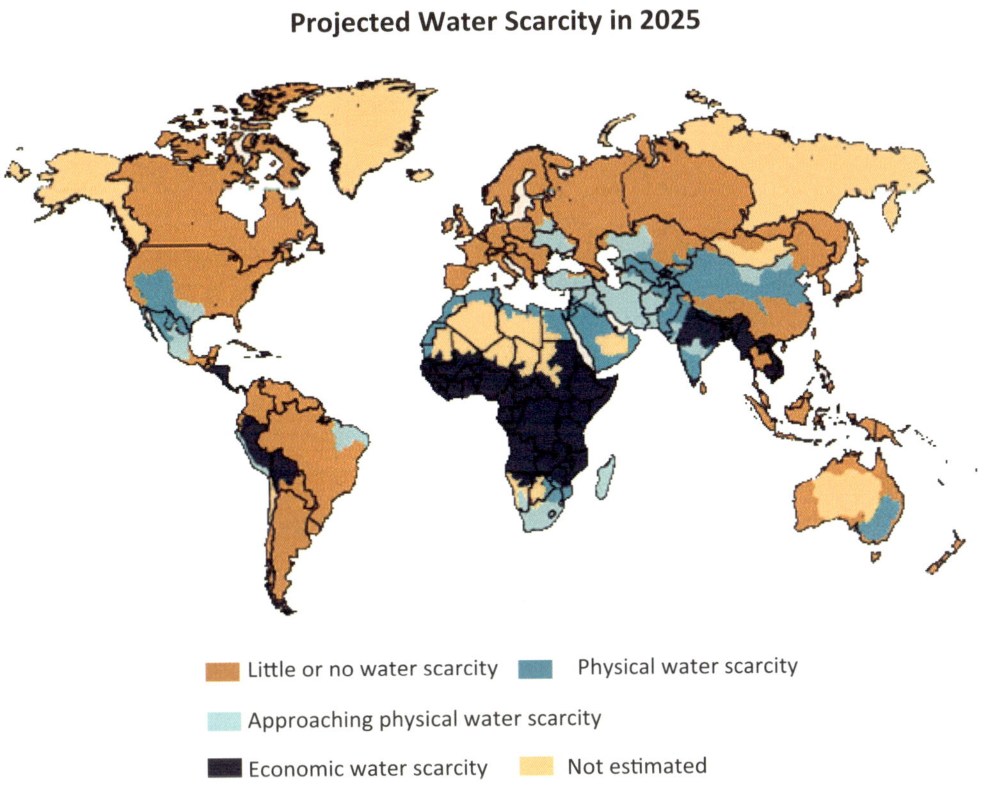

a. Which countries will face physical scarcity of water?
b. Which regions will face the economic scarcity of water?
c. Does water scarcity have an impact on human health? If so, what are some of the consequences?
d. Give three reasons why some nations are experiencing a scarcity of clean drinking water.
e. What would you predict the climate to be in these areas and why?
f. Predict the difference in per capita income (average income per person) between regions with plenty of water and areas with economic water scarcity.

Water Scarcity in India

A NITI Aayog report in 2018 stated that 600 million people, or nearly half of India's population, face extreme water stress. Three-fourths of India's rural households do not have piped, potable water and rely on sources that pose a severe health risk. India has become the world's largest extractor of groundwater, accounting for 25 per cent of the total, and 70 per cent of our sources are contaminated. Our major rivers are dying because of pollution. It concluded that 'India is suffering from the worst water crisis in its history'.

Case Study

Analyse the infographic shown here.

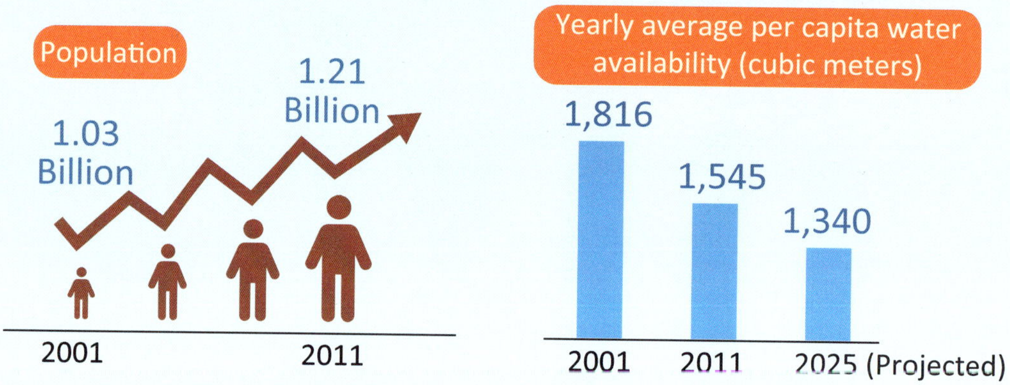

a. What is the leading cause of water scarcity in India?

b. What is the difference between the projected per capita water availability between the years 2001 and 2025?

c. Give two reasons for the difference in water availability over the years.

d. Write two things that you can do in your community to reduce scarcity.

Reflect

'India is suffering from the worst water crisis in its history'. Do you agree with this statement? What can you do to help?

Consequences of Water Scarcity in Our Country

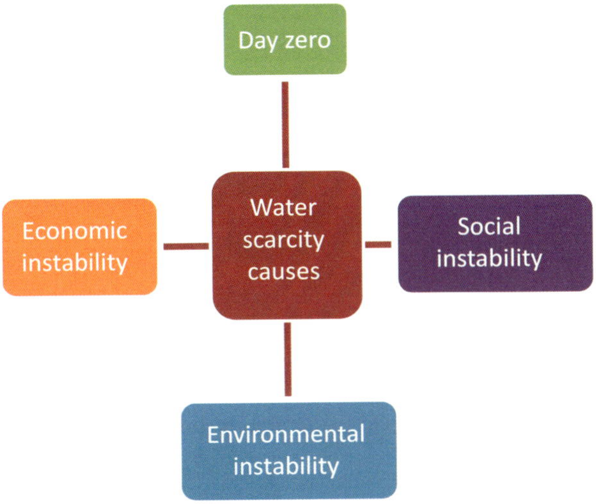

Day Zero: The day when a city's taps dry out and people have to stand in line to collect a daily quota of water. According to the 2018 WaterAid report, 12% of India's population is already living the Day Zero scenario, with no access to clean water near their homes forced to dig deeper wells or buy water.

Dealing with Water Scarcity: Saving and Reclaiming Water

There is an urgent need to focus on water conservation measures to counter the effects of increasing water demand and overuse of groundwater. The only way to deal with this crisis is by making immediate changes in how we think about environmental responsibility and our lifestyle. In India, all water bodies are not only of great religious importance but also help in building international relations and support the economy. The public health and safety are also interconnected through water resources.

It is easier for Indians to adopt this lifestyle because sustainability is rooted in our culture and heritage. Traditionally, we have always focused on rainwater sources for our water consumption. There are many traditional methods that are followed across the country to harvest rainwater. The ancient forts architecture supported elaborates a rainwater harvesting system. However, with modern technology we now access groundwater that is deep inside the ground. Our present society is mainly using groundwater because we have almost exploited other water sources and have forgotten the practice of conserving rainwater.

Saving Water: A traditional yet modern approach

Rainwater harvesting (RWH) is one of the best solutions to combat water scarcity problems all over the country. First of all, it is important to realize that RWH is not a new phenomenon in India. It has been there for centuries as a primary source of irrigation. Along with this, there are many traditional and modern innovative water conservation methods in the Indian culture. In today's scenario, it is essential to reintroduce these methods into our infrastructure, bringing back the conventional eco-friendly practices that are no longer popular.

In urban areas, buildings are usually constructed with rooftops made of reinforced cement concrete (RCC), Mangalore tiles, asbestos/galvanized iron/zinc sheets, etc. Usage of such materials in construction of a building requires removal of rooftop rainwater from the structure, leading to letting off the water into storm water drains outside the plot area, which eventually goes away from the city.

Furthermore, as such rooftops are built with impervious materials, let rainwater runoff in large quantities, and the loss due to evaporation and percolation is minimal, these can be used for rooftop rainwater harvesting. By using storage devices such as masonry tanks/ferro cement tanks/plastic or metal containers on (a) roof, (b) ground level or, (c) below the ground rooftop water can be stored.

When made with RCC, flat roofs usually have a waterproofing on the surface. The waterproof system is created with a slight slope towards the downtake pipes. These days a rich cement mortar is used on rooftops along with weatherproof tiles laid on it. These roofs are ideal for rooftop RWH.

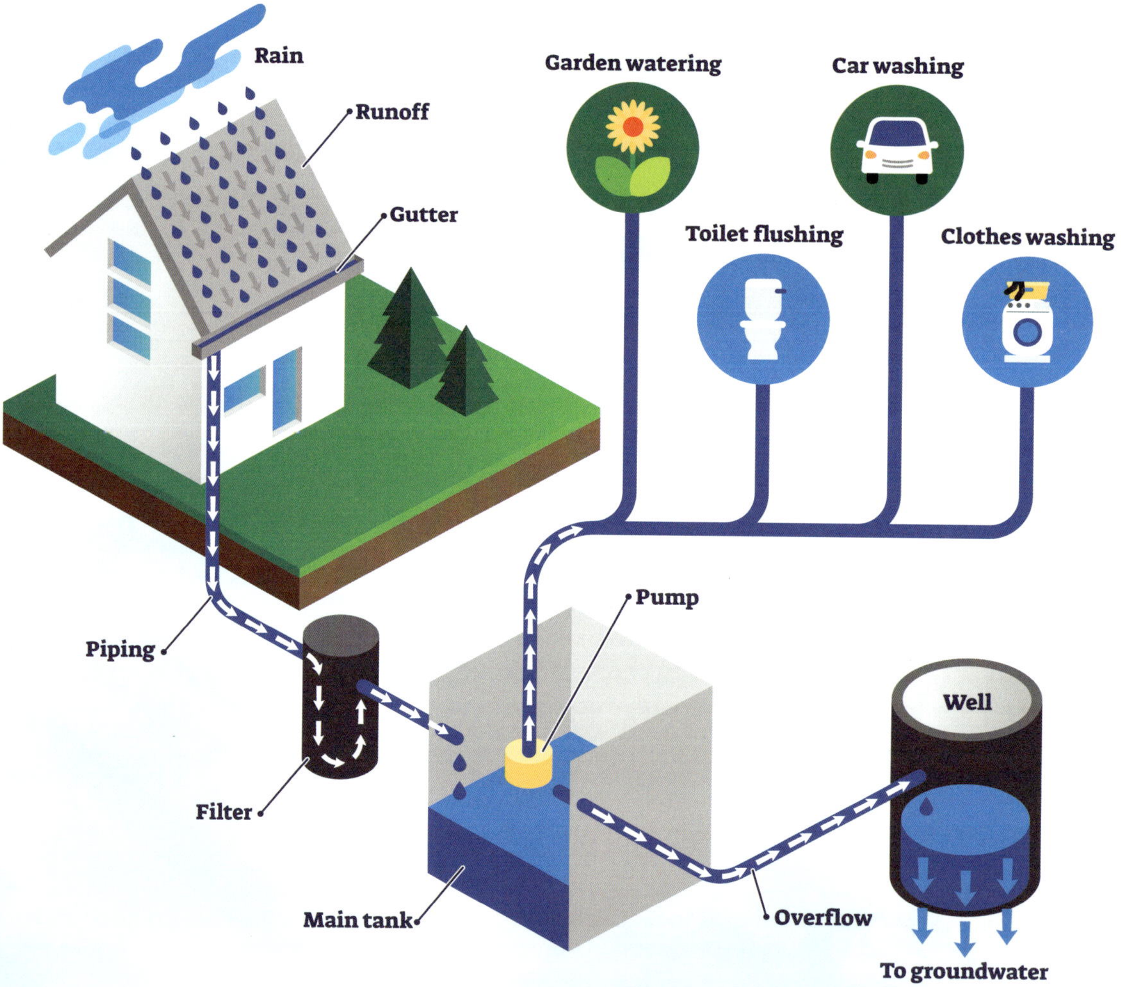

Activity

1. Check out your school building and houses in your residential area. Do they have RWH provisions or are most of them connected to storm water drainage?
2. Devise a simple strategy for your school or community to improve RWH.
3. In the following figure, designs of houses located in various parts of India are shown.

   a. How will you improve the designs so that rainwater is harvested?
   b. What other structure/equipment will you add so that the water is usable?
   c. Which types of roofs are best for rainwater harvesting? And, Why?

The Government of India, under Jal Shakti Abhiyan launched a programme titled 'Catch the Rain'. The programme has a state-specific approach. For your state, find out a plan of action suggested under Catch the Rain Programme.

TASK

Your school has been nominated to launch a digital awareness campaign for 'Catch the Rain' programme. Make a video/audio/social media page to highlight the importance of saving rainwater and how we can do it individually at our homes. Post your entry on Facebook page. Use QR code as on page no: 44

Chapter 5

Wastewater at Homes

The water we use in our everyday chores is clean water sourced from ground or surface water. When used at home, two types of wastewater are generated using the clean water: grey water and black water.

TYPES OF DOMESTIC WASTEWATER

GREY WATER

This contains light and dark grey waters. Light grey water is the domestic wastewater that can be reused for some purpose without treatment, e.g. drinking water RO reject. Dark grey water is generated from the kitchen, bathroom (bathing, handwashing), and laundry activities. It has a low level of contamination to be treated easily and reused in gardens or lawns.

BLACK WATER

Black water comes from toilets and urinals. It contains organic matter, nutrients (nitrogen and phosphorous), and a high microbial load. It, therefore, needs a high level of treatment before discharge or reuse.

Activity

How Much Water Do We Use?

We have the luxury of getting water by turning on taps. We use water for almost everything. Have you ever thought how much water actually your family uses every day?

Know your water meter: Find a water meter at your home or at your school. It is usually placed at the front of the property. In case your school has more than one meter, you should add readings of all the meters together. Your water meter looks something like the one given in the picture. On the front of the meter, you would see a series of numbers. The numbers measure the amount of water used in kilolitres and litres. On most water meters, the black numbers measure kilolitres and the red numbers measure litres.

Digital display meters: These meters have six, seven, or eight digits displayed.

If your meter looks like the following one, then kilolitres are shown by the black digital display and litres by the dials. The dials can be read from right to left.

When a pointer rests between two numbers, use the lower number.

Dial Face display meters

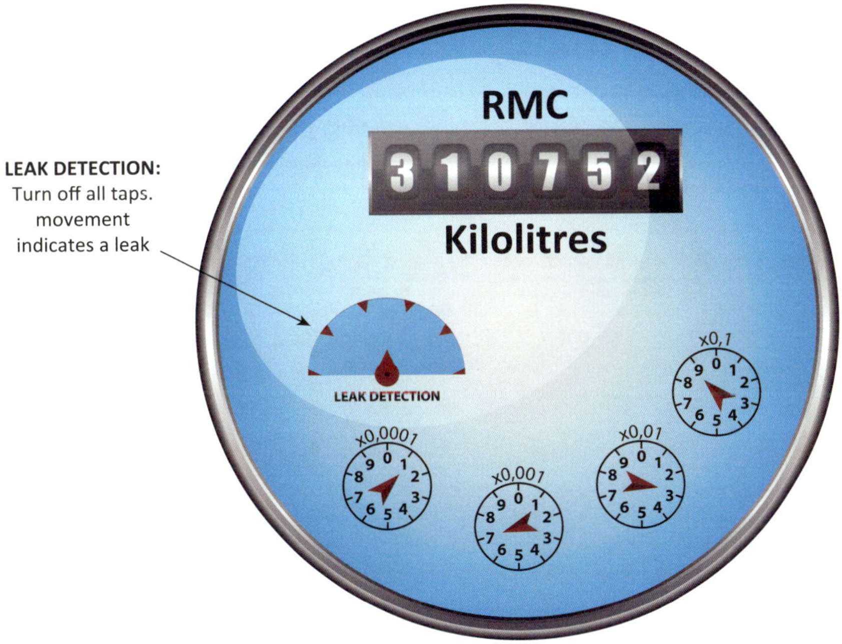

Activity

How grey is my water?

1. Note your meter's reading every day for 10 days.
2. Record the number of buckets you use to take a bath. Based on the size of the bucket, calculate the amount of water you use to take a bath. Calculate the average amount of water used by your family in these 10 days.
3. Record the number of times the flush in your bathroom is used in a day. Based on the size of the flush tank, calculate the amount of water used in a day for flushing. Calculate the average amount of water being used by your family.
4. Based on the above data, analyze the amount of black and grey water your family generates.
5. List two ways in which we can use grey water.

Chapter 6

Being Water Wise: Using and Reclaiming Wastewater

Wastewater must no longer be seen as a problem but as a solution, which can help provide sustainable infrastructure services, improve financial viability of operators and environmental quality, and strengthen the resilience of the systems. Wastewater around the world is seen as a burden. However, it should be treated as a resource.

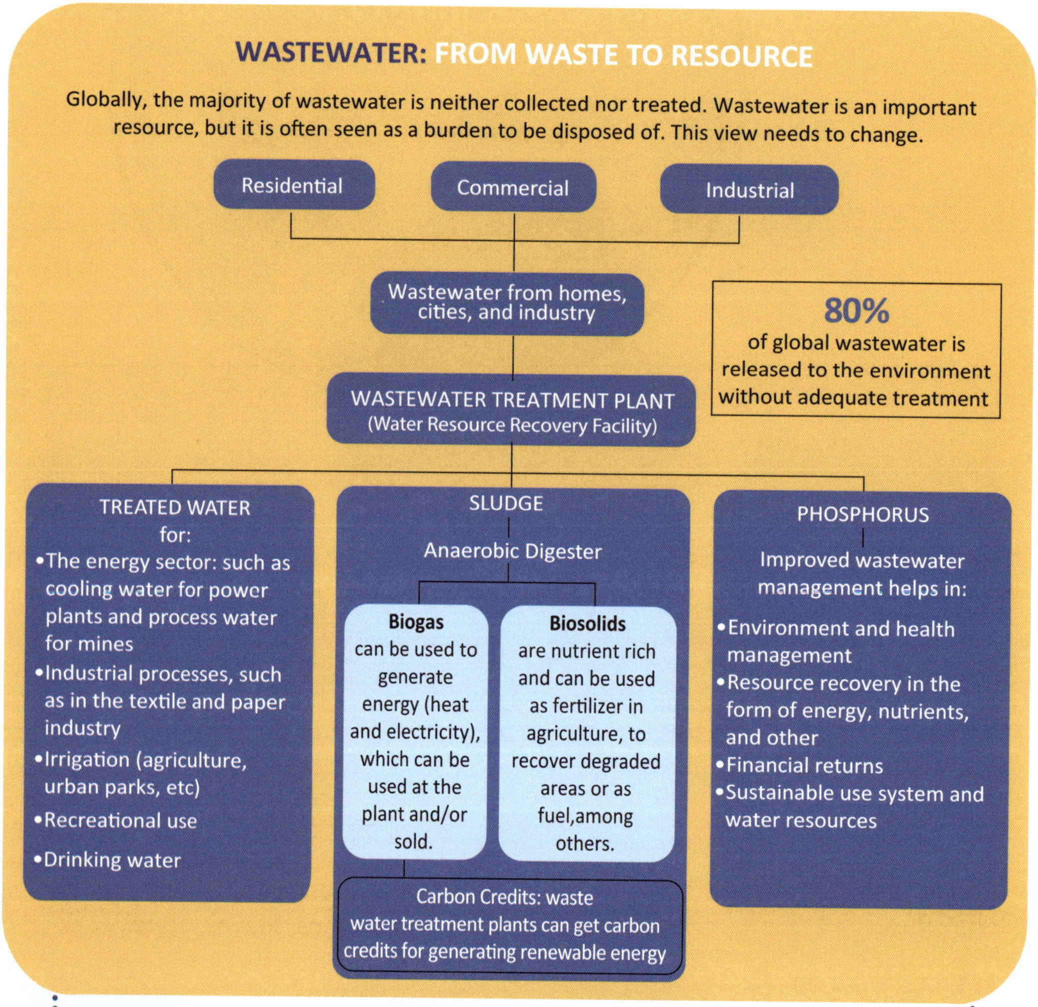

Small Steps that Can Help in Solving Water Problems

Using light grey water

Light grey water is the domestic wastewater, which can be reused for some purposes without treatment, for example, RO reject drinking water.

Reuse of light grey water and grey water to reduce freshwater requirement

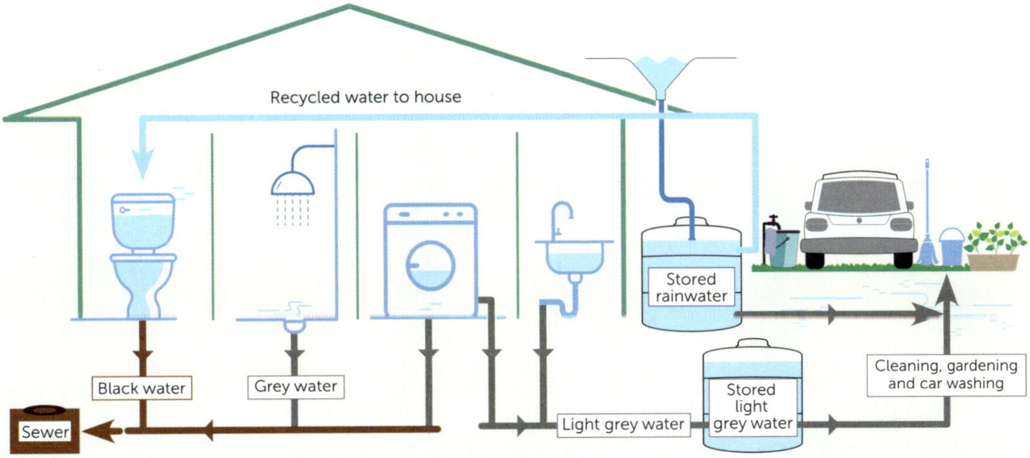

Grey water reclamation is the process by which households use grey water's potential instead of simply piping it into overburdened sewage systems with all the black water.

Each wastewater type must be treated differently and can be used in various ways. Grey water is ideal for garden watering, with appropriate precautions, such as using low or no sodium and phosphorus products and applying the water below the surface. Appropriately treated grey water can also be reused indoors for toilet flushing and clothes washing, which are both significant water consumers. Grey water recycling helps in reducing household water usage by about 50% – potentially halving your water bills as well as your water footprint.

Water after washing vegetables, fruits, pulses, etc. | **Last rinse off washing machine** | **Aquarium water** | **AC, RO run-down**

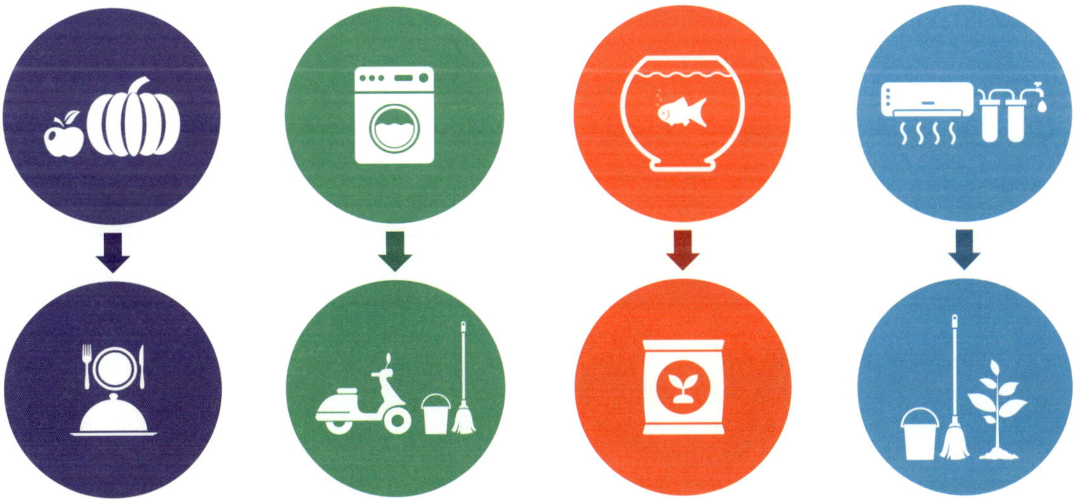

Rinse dishes | **Clean vehicles · Floor** | **Fertilizers for plants** | **Multi-purpose**

Case Study

Grey water usage in Indian cities

Analyse the following graph and answer the questions that follow:

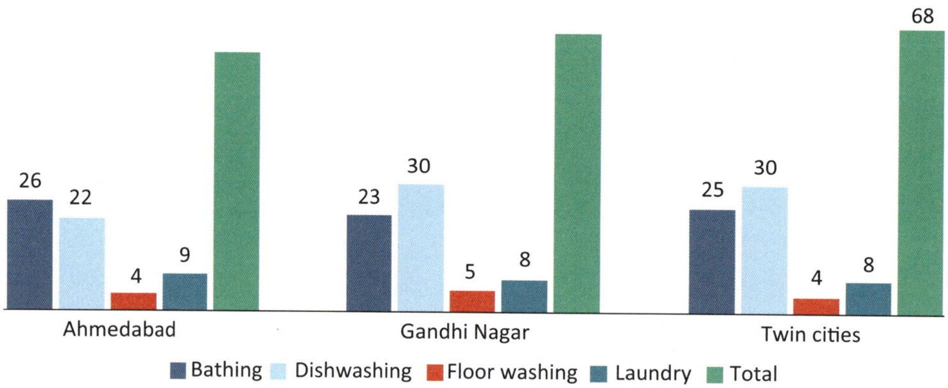

https://timesofindia.indiatimes.com/city/ahmedabad/gnagar-uses-26-more-water-per-capita-than-abad/articleshow/82122200.cms

a. Which city uses more water for bathing?

b. Which city shows more usage of water?

c. If you have to plan grey water usage for these cities. What suggestions and ideas will you give?

TASK

Design a water-saving system for your bathroom that reduces water consumption and wastage.

Cities Across the Globe that Reuse their Wastewater

Given here are some examples of cities that recycled water to meet their growing needs of water. Read how they use their wastewater. Analyse the use of wastewater in these cities.

San Diego, California

San Diego's Water Purification Demonstration Project in 2012 showed how wastewater could be turned into a 'reliable, sustainable and diversified' local purified water supply, suitable for drinking. The public approval for water reuse has also increased from 26% in 2004 to 73% in 2012. The project, now known as Pure Water San Diego, was a 2015 winner of the US Water Alliance's US Water Prize, and targets supplying the city with 83 million gallons of drinking water locally by 2035.

The Merlion City, Singapore

Singapore is another beacon of wastewater success. Its NEWater initiative uses a process of microfiltration, reverse osmosis, ultraviolet disinfection, and alkaline 'pH' balancing, to supply 40% of the city-state's current water needs for both ultraclean and potable purposes from five water treatment plants. It aims to increase this to 55% by 2060.

Aqaba, Jordan

The city of Aqaba in Jordan, one of the world's most water-scarce countries, collects and treats 90% of its wastewater – around 31,000 million m^3/d. The city's resource recovery strategy has helped to reduce its carbon emissions, green its urban landscapes, support both tourism and public health, and generate more than US$ 4 million for the Aqaba Water Company (AWC). Meanwhile, new legislation in Manilla, the capital of the Philippines, commits all stakeholders to manage and reuse 100% of their wastewater by 2028.

Europe's Action on Reusing Its Wastewater

In 2020, the EU made a big step forward towards a consistent, coherent pan-EU water-reuse framework when the European Parliament approved the Water Reuse Regulation. The new law defines minimum requirements at European level for the first time for reclaimed water to be used for agricultural purposes in a safe way, protecting people and the environment. The new regulation is also aimed at reducing the over exploitation of aquifers, considered one of the biggest threats to Europe's water resources.

Reflect

Based on the information provided on reuse of wastewater in different cities, reflect and complete the following:

1. Summarize your opinion about wastewater reuse as an approach to combat water scarcity. Do you think it's safe?
2. As president of Residents Welfare Association of your community, ask the residents about their views on water reuse. If they find the idea objectionable, find out why, and see if you can convince them otherwise. Prepare a report on your conversation by briefly summarizing 4-5 critical aspects of discussion.
3. In light of your discussion with your friends or family, summarize whether you were able to change their views on reuse and reclamation of wastewater.

TASK

Life cycle of a Water Droplet

Develop a comparative story of two water droplet brothers, Ram and Sham. They narrate their life story wherein one brother is treated and recycled after usage whereas the other is not.

Chapter 7

India's Policies, Perspectives, and Initiatives for Reusing Wastewater

For a nation-like India, water is an essential resource as its economy is heavily dependent on agriculture and agriculture needs water. Also, our river system supports the ever-growing population of the country. Water bodies are not only receding, they are getting polluted at a faster rate than ever due to our callousness and indifference towards resources and environment. Therefore, to protect water and keep it clean several laws and policies have been formulated by the Government of India.

The timeline here highlights the various policies formulated by the Government of India to conserve, reuse, recycle, and save water.

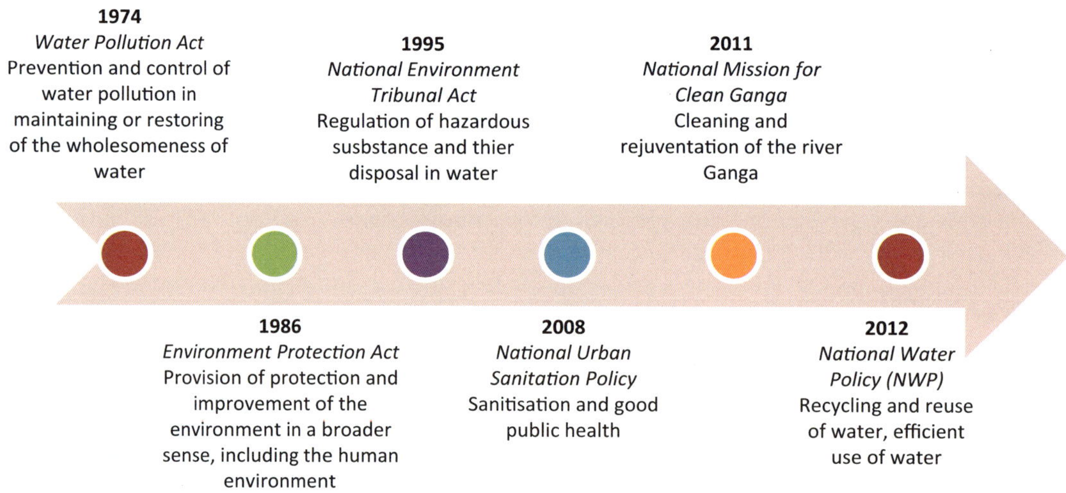

The increasing water scarcity the country is facing in major cities including Chennai, Bangalore, Delhi, and Hyderabad points to the need to focus efforts on wastewater reuse, which falls under the purview of sanitation efforts. India treats less than 30% of all faecal sludge – a mix of black water and excreta – generated. And even the little that does get treated is then discharged into open drains or water bodies and thus not being effectively reused.

Recycled water can satisfy most water demands as long as it is adequately treated to ensure the water quality is appropriate for the use it is intended for. For example, recycled water can be used for non-potable purposes such as landscaping, watering public parks, and construction activities. It can also be used in industries such as textiles, power plants, paper mills, and oil refineries.

The above-mentioned ways of reusing wastewater would drastically reduce undue stress on our water tables.

The Jal Shakti Abhiyan is a large-scale water conservation effort with a five-fold focus: water conservation and rainwater harvesting, renovation of traditional and other water bodies, reuse of water and recharging of structures, watershed development, and intensive afforestation.

The table given here outlines the role of Jal Shakti Abhiyan.

Ministry of Jal Shakti	Formed in May 2019 by merging Ministry of Drinking Water and Sanitation, Ministry of Water Resources, River development and Ganga Rejuvenation for optimal sustainable development, maintenance of quality and efficient use of water resources.	1. Overall planning, policy formulation, coordination and guidance for water resources 2. Technical guidance, scrutiny, clearance and monitoring of the irrigation, flood control and multi-purpose projects 3. General infrastructural, technical and research support for development 4. Providing special central financial assistance for specific projects 5. Overall policy formulation, planning and guidance in respect of irrigation management 6. Overall planning for the development of groundwater resources 7. Formulation of national water development perspective 8. Coordination, mediation, and facilitation of interstate interests 9. Operation of the central network for flood forecasting 10. Inter-state negotiations 11. Ensure effective abatement of pollution and rejuvenation of the river Ganga by the Ministry of Jal Shakti, 2019

TASK

In the Ministry of Jal Shakti Abhiyan, you are appointed a member of water regulatory authority of your community. What steps will you take in your community while taking the Jal Shakti Abhiyaan in account? Write at least five steps you will take to reuse and reclaim wastewater in your community.

Please share it on Facebook page. Use QR code as on page no: 44

Chapter 8

Goal 6 for India: Work towards United Nations Sustainable Development Goals

The Sustainable Development Goals have committed the international community to expand international cooperation and capacity building on water- and sanitation-related activities and programmes, and support local communities in improving water and sanitation management. Through Goal 6, the countries have resolved to achieve universal access to safe drinking water and adequate sanitation and hygiene to all in the next 15 years.

SDG 6 deals with all aspects of water availability, access, and use, and calls upon all nations to 'Ensure availability and sustainable management of water and sanitation for all'. The eight targets of SDG 6 are as follows:

- 6.1: Provision of drinking water
- 6.2: Sanitation and hygiene services
- 6.3: Treatment and reuse of wastewater and ambient water quality
- 6.4: Water-use efficiency and scarcity
- 6.5: Integrated water resources management
- 6.6: Protecting and restoring water-related ecosystems
- 6.7: International cooperation and capacity-building
- 6.8: Local participation in water and sanitation management

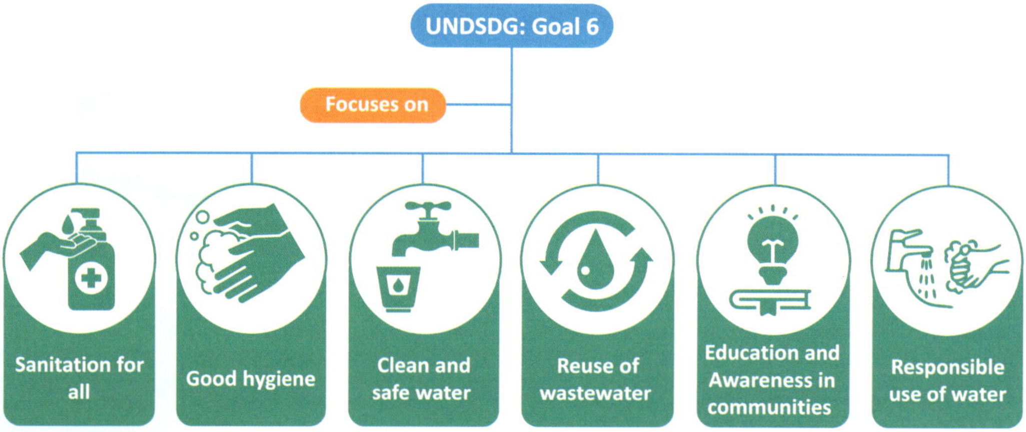

Activity

Link: https://www.margreetdeheer.com/the-global-goals-of-sustainable-development/

Follow the link given here and examine the cartoon for UNSDG Goal: 6. For your social media page, you are developing an animation to support UNSDGs. Write a script using for animation based on this cartoon (using at least three elements) to generate awarness about UN SDG 6.

India and Goal 6

The overall proportion of Indian households with access to improved water sources increased from 68% in 1992–93 to 89.9% in 2015–16. However, in 2015–16, 63.3% of rural households and 19.7% of urban households were not using improved sanitation facilities. According to the World Bank, more than 520 million in India were defecating in the open – the highest number in the world. This figure is expected to have reduced significantly given that improving sanitation is a key priority of the central government. It has introduced several flagship programmes including the Swachh Bharat Abhiyan to clean India, the National Rural Drinking Water Programme, and Namami Gange that aims at the conservation of River Ganga.

Targets to be Achieved by 2030

- Universal and equitable access to safe and affordable drinking water for all.
- Access to adequate and equitable sanitation and hygiene for all and end open defecation, paying special attention to the needs of women and girls and those in vulnerable situations.
- Improve water quality by reducing pollution, eliminating dumping and minimizing release of hazardous chemicals and materials, halving the proportion of untreated wastewater and substantially increasing recycling and safe reuse globally.
- Substantially increase water-use efficiency across all sectors and ensure sustainable withdrawals and supply of freshwater to address water scarcity and substantially reduce the number of people suffering from water scarcity.
- Implement integrated water resources management at all levels, including through transboundary cooperation as appropriate.
- Expand international cooperation and capacity-building support to developing countries in water- and sanitation-related activities and programmes, including water harvesting, desalination, water efficiency, waste water treatment, recycling and reuse technologies.

One of the main focuses of Goal 6 is to provide clean water to people. In India, as per the latest NSSO report, only 56% of urban households have access to piped water in their houses or plots and close to 33% of urban households still rely on public taps or tube/bore wells. The public taps, tube, and bore well are often contaminated with faecal matter and other contaminants due to open defecation. The rapid pace of urbanization, which is growing at an annual growth rate of 2.1%, is further adding to the problem and sustainability of water and sanitation services are expected to be affected even more severely in the future.

Analyse the infograph given here.

GLOBALLY

2.6 BILLION
PEOPLE HAVE GAINED ACCESS TO IMPROVED DRINKING WATER SOURCES SINCE 1990

666 MILLION
PEOPLE ARE STILL WITHOUT WATER

IN INDIA

NEARLY **18%** OF WORLD'S POPULATION BUT ONLY **4%** OF AVERAGE GLOBAL RUNOFF IN RIVERS

NEARLY **500 MILLION** PEOPLE ARE AFFECTED BY DROUGHT IN INDIA

OVER **20%** OF THE POPULATION LIVES IN STATES WHICH ARE NOT DEFECATION FREE

1/5 CHILD DEATHS DUE TO SEVERE DIARRHEA ARE IN INDIA

EACH YEAR NEARLY **102,813** CHILDREN DIE DUE TO SEVERE DIARRHEA

a. What challenges are faced by India to achieve Goal 6?
b. How does open defecation impact water supply?
c. Devise a small plan to provide clean water in a slum area near your community. The plan should be cost-effective and innovative. Develop a sales pitch for your plan.

Activity

Download the SDG index report. Collect the data for SDG 6 for your state and three neighbouring states. Compare the performance of the states in the area of clean water and sanitation.

As an aware citizen, what actions will you take to contribute to the goals of SDG 6? Make a short video describing the contribution you would make.

Chapter 9

My Water Goals: Everyone Can Make a Difference

Clean, freshwater is essential to life. You can help bring clean water by reclaiming wastewater to tackle water scarcity across many parts of the world. We often think that one person's action cannot make a difference in the world. The truth is individual actions can also bring about substantial change in the community.

Let us look at some individuals who have brought this change in their community and are making a difference in the world.

Noida, India

Ramveer Tanwar, from Dadha village in Greater Noida, has been reviving lakes and ponds since 2014. Single-handedly, through his efforts, he brought 10 ponds in Greater Noida in Uttar Pradesh to life again.

He used a simple device consisting of a simple double filtration system so that any water entering the lakes would have to pass through a mesh of wooden planks and then a mesh of grass. The mesh of wooden planks helped stop more oversized items like plastic bottles, and the subsequent mesh then further filters the incoming water at the next stage. Many of the villagers are coming forward; some joined hands to do the hard labour and some provided Mr Tanwar with helpful tools and equipment in this process of cleaning and reviving ponds and lakes.

He has been awarded several awards like Rashtriya Swayamsidh Samaan, Rex Karamveer Chakra Award, Shining World Protection Award for his commendable work in water conservation.

California, USA

In California, teenager ***Shreya Ramachandran*** noticed the harmful effects of droughts, so she designed a system for households to reuse grey water. Instead of regular soap, Shreya used soap nuts or reetha, a plant that is often used to clean clothes in India.

She found that grey water from soap nuts, as well as several organic detergents, could be reused safely for non-potable uses. But grey water that was generated from conventional- or chemical- based soaps have soluble salts and boron which accumulated in the grey water and makes it unusable for crop irrigation.

She went on to found her own nonprofit organization, The Grey Water Project. It teaches people how to recycle grey water in their own homes. She does workshops at schools, libraries and corporate events, and has developed a grey water science curriculum that has been implemented in more than 90 schools so far.

Her ultimate goal is essentially for grey water reuse to be just as common as paper or plastic recycling.

TASK

Share your views on how individual efforts and small communities can bring about change. Focus your discussion on how sustainable habits make the world a better living place. Please share it on Facebook page. Use QR code as on page no 44.

Capstone Activities

- Based on your learnings in this programme, what immediate actions will you take at home to save, conserve, and reuse wastewater?
- Design a programme for your community based on sustainable practices of living that can help in reusing the wastewater (grey, dark grey, and black).
- Create a social media page to inspire children, youth, and adults to reuse or reclaim the used water.

Use this QR code to view the project Facebook page on your/ your parents' mobile phone.